SPACE FIRSTS™

SPUTNIK
The First Satellite

Heather Feldman

The Rosen Publishing Group's
PowerKids Press™
New York

For Katie Dinner and Sydney Weber—may all your dreams come true

Published in 2003 by The Rosen Publishing Group, Inc.
29 East 21st Street, New York, NY 10010

First Edition

Edited by: Nancy MacDonell Smith
Designed by: Mike Donnellan

Photo Credits:Cover, pp. 7, 12, 15, 16, 19 © Photri-Microstock, Inc.; pp. 4, 11 © Bettmann/Corbis; p. 8 © Corbis; p. 15 inset © Hulton/Archive/Getty Images; p. 20 © Reuters NewMedia Inc./Corbis.

Feldman, Heather.
Sputnik : the first satellite / by Heather Feldman.— 1st ed.
    p. cm. — (Space firsts)
Includes bibliographical references and index.
ISBN 0-8239-6244-X (lib. bdg.)
1. Sputnik satellites—History—Juvenile literature. 2. Artificial satellites—Soviet Union—History—Juvenile literature. 3. Artificial satellites, Russian—History—Juvenile literature. [1. Sputnik satellites. 2. Artificial satellites, Russian. 3. Rocketry.] I. Title.
TL796.5.S652 S6644 2003
629.46'0947—dc21

2001006024

Manufactured in the United States of America

Contents

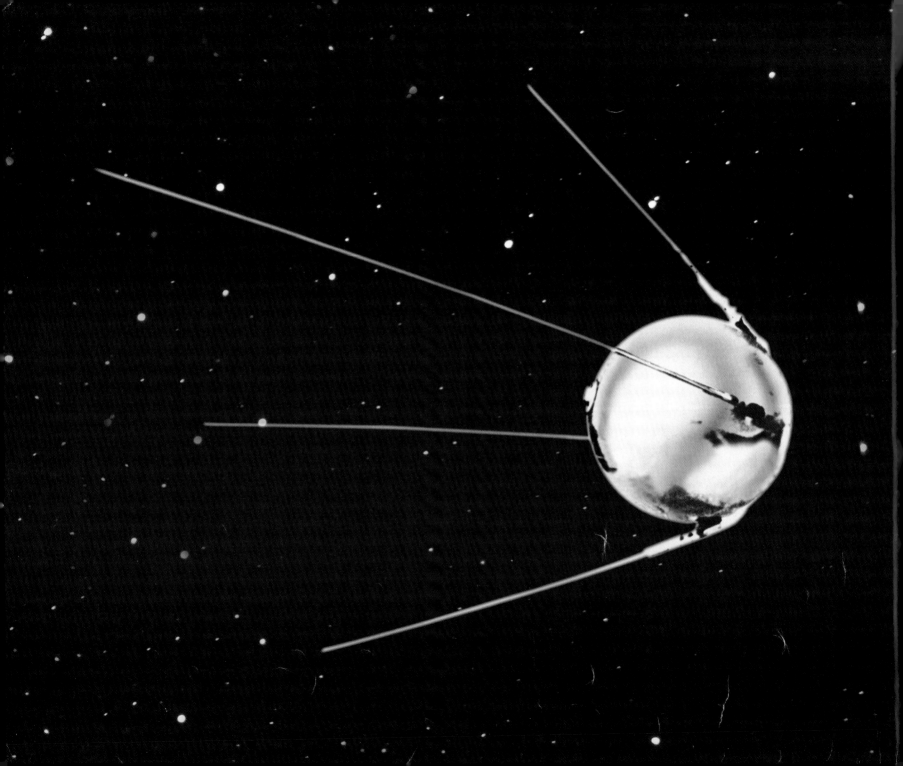

The Space Race

The space race was a contest between the Soviet Union, which was sometimes called Russia, and the United States to see who could get into space first. Both countries wanted to be the strongest and the most powerful country in the world.

In the early days of the space race, Russia and the United States focused on using rockets for military purposes. They wanted to make sure their rockets could be used to fight their enemies. The rockets had to be powerful enough to hit targets that were very far away. After a while, both countries began to focus on making scientific discoveries and achieving firsts in space. It was a race to see who could become the **superpower** of the world. Russia took the lead with the **launch** of *Sputnik 1*, the first **satellite**.

The Russian satellite Sputnik 1 *was the first human-made object to circle Earth. Its journey lasted for three months.*

Russia's First Win

On October 4, 1957, *Sputnik* was launched from Kazakhstan, a state in the Soviet Union. This was the first successful space launch in world history. The Russians had been working very hard on their space program and had kept it a secret from the rest of the world. The launch of *Sputnik* surprised everyone. Russia proved that it was both a powerful military force and a scientific leader.

The full name of the satellite was *Sputnik Zemli*. This means "fellow traveler of Earth." In the United States, this famous satellite was better known as simply *Sputnik*. Once *Sputnik* was launched, the space race became more serious. *Sputnik* led to many more achievements in space by both powerful countries.

Sputnik's rockets were bigger than the satellite. Sputnik's launch was announced on October 5, 1957. The space age had begun!

What Is a Satellite?

Sputnik was a satellite. A satellite is an object that **orbits** Earth or another object in space. An orbit is a path that a satellite follows. Satellites are launched into space to collect information about our vast universe. Satellites study things such as the temperature in space and the different gases that make up the **atmosphere** in space. Satellites also collect information on the rays that stars and galaxies give off, such as **X rays** and **ultraviolet rays**.

Before satellites, **astronomers** used telescopes to study space. Those telescopes did not give a very clear view of space. The atmosphere around Earth is full of dust and moisture, which makes space difficult to view with a telescope. Satellites travel beyond this atmosphere and give scientists a much clearer view of space.

Early telescopes such as this one, which was made in 1873 for the U.S. Naval Observatory, were not powerful enough to see very far into space.

Launching a Satellite

It takes a lot of power to launch a satellite into space because Earth's **gravity** is very strong. A satellite, such as *Sputnik*, travels at very high speeds. Satellites can go as fast as 17,400 miles per hour (28,003 km/h). Rockets are powerful enough to travel this fast because they carry liquid fuel and the liquid oxygen needed to burn this fuel.

There is no oxygen in space, so rockets have to carry their own supply. The gas produced from the burning liquid fuel is shot out of the rocket. As long as the fire burns and the gas is released, the rocket will fly. Even so it takes several rockets working together to create enough power to launch a satellite into space.

Explorer 1, *the first American satellite to be successfully launched into space, was powered by three rockets.*

Early Rockets

As many as 1,000 years ago, the Chinese were the first to use rockets. These early rockets were like fireworks, but they were used as weapons. When the gunpowder inside was lit, the rocket thrust forward toward the enemy. In the 1880s, a Russian man named Konstantin Tsiolokovsky began studying rocket spaceflight. He figured out that a rocket needed to travel close to 5 miles per second (8 km/s) to stay in the air. Tsiolokovsky also came up with the idea of mixing liquid fuel and liquid oxygen to power rockets.

In 1926, American scientist Robert Goddard designed and launched the first liquid-fuel rocket into space. It stayed in the air for only 2 ½ seconds. Soon scientists designed rockets that would be powerful enough to escape Earth's atmosphere.

This photo of Robert Goddard was taken in 1935. Inset: This is Goddard in 1915, with an early model of his famous 1926 rocket.

Facts About Sputnik

The Russian satellite *Sputnik* was an **aluminum sphere** about the size of a beach ball. It weighed 183 pounds (83 kg) and took about 96 minutes to orbit Earth. *Sputnik* sent out a signal that sounded like "bleep, bleep." *Sputnik* traveled around Earth collecting information for 21 days.

The satellite had two radio transmitters and a battery. It also contained instruments used to measure the conditions in space. These instruments radioed data to Russian scientists about **cosmic rays, meteoroids**, and the **density** of the upper atmosphere. When *Sputnik*'s journey was finished, it reentered Earth's atmosphere and was destroyed by intense heat, as scientists had expected. The mission was a success!

Sputnik was no bigger than a beach ball, but it was much larger than the first American satellite, which was the size of a carrot. Inset: *This photo of Sputnik was taken just before it was launched.*

The World Reacts to Sputnik

The launching of *Sputnik* was a tremendous achievement for the Russians. No one could have predicted the impact *Sputnik* would have on the world. *Sputnik* proved that people can actually overcome the problem of gravity and can reach space. For military leaders, *Sputnik* was important because it showed that powerful missiles can travel between continents at extremely fast speeds. This small satellite truly changed the world and sparked a new enthusiasm in people everywhere for space programs and for advancements in technology.

Since the launch of Sputnik, satellites have become very common. Satellites such as this one have helped us to learn more about our universe.

America's Reaction to Sputnik

Sputnik did a lot for American science and technology, too. In order to compete with Russia, America knew it had to get its space program up and running quickly. One year after *Sputnik* was launched, the United States created the National Aeronautics and Space Administration (NASA) and a space program to match the Russian space program. *Sputnik* also caused a heated national debate about the quality of science education in America. Some people thought Americans needed to study more math and technology.

On January 31, 1958, *Explorer 1*, the first American satellite, was launched into space. This satellite orbited Earth for 12 years.

About four months after Sputnik was launched, the United States launched Explorer 1, which was shaped like a pencil and weighed 29 pounds (13 kg).

Modern-Day Satellites

Sputnik was the first step toward the very advanced satellites used around the world today. Satellites continue to give astronomers new and exciting information about what Earth and our universe are really like. For example, in 1983, a satellite called the **Infrared** *Astronomical Satellite* (IRAS) was launched. This satellite discovered new **comets**.

Satellites have also helped in our global communications. This means satellites help to send television pictures to our living rooms. Some satellites are used in the **navigation** of ships and airplanes. Weather satellites help us to predict and prepare for bad weather. With so many satellites in space right now, some scientists worry that space will soon become overcrowded with satellites!

This is NASA's Compton Gamma Ray Observatory satellite. This bus-size satellite was in orbit around Earth from April 1991 to June 2000.

The Future of Space Exploration

Scientific discovery in space continues to thrive today. The space race has changed a great deal. It is no longer a contest. Today American and Russian scientists work together to explore new frontiers in space. Space research is not so much about competition, it is more about cooperation. Scientists are using the research gathered in the past to plan for our future in space. American and Russian astronauts now spend time together in space stations gathering information and making new discoveries. We have come a long way since *Sputnik* was launched. There is most certainly more to come in our quest to learn about space.

Glossary

aluminum (uh-LOO-mih-num) A type of metal.

astronomers (uh-STRAH-nuh-merz) Scientists who study the Sun, moons, planets, and stars.

atmosphere (AT-muh-sfeer) The layer of gases that surrounds an object in space.

comets (KAH-mits) Heavenly bodies, made up of ice and dust, that look like stars with tails of light.

cosmic rays (KAHZ-mik RAYZ) High-energy particles that come from outside the solar system.

density (DEN-sih-tee) The heaviness of an object compared to its size.

gravity (GRA-vih-tee) The natural force that causes objects to move or tend to move toward the center of Earth.

infrared (in-fruh-RED) Electromagnetic radiation with long wavelengths that is found in the invisible part of the spectrum.

launch (LAWNCH) When a spacecraft is pushed into the air.

meteoroids (MEE-tee-uh-roydz) Small objects, often pieces of comets, traveling through space.

navigation (na-vih-GAY-shun) The act of steering, directing, or sailing a ship or aircraft.

orbits (OR-bits) Travels in a circular path around an object in space.

satellite (SA-til-eyet) A human-made or natural object that orbits another object.

sphere (SFEER) An object that is shaped like a ball.

superpower (SOO-per-pow-er) A nation, like the United States or the Soviet Union, that dominates world affairs because of its military strength.

ultraviolet rays (ul-trah-VEYE-oh-let RAYZ) Rays given off by the Sun that are dangerous to our skin and eyes.

X rays (EKS RAYZ) A form of electromagnetic radiation, similar to light but of shorter wavelength and capable of penetrating solids.

Index

A

astronauts, 22
astronomers, 9, 21
atmosphere, 9, 13–14

C

comets, 21
cosmic rays, 14

E

Earth, 9, 13–14, 18, 21
Explorer 1, 18

G

Goddard, Robert, 13
gravity, 10, 17

I

*Infrared Astronomical
Satellite* (IRAS), 21

L

liquid oxygen, 10, 13

M

meteoroids, 14
missiles, 17

N

National Aeronautics and
Space Administration
(NASA), 18

R

rockets, 5, 10, 13
Russia, 5–6, 18

S

satellite(s), 5–6, 9–10, 14,
17–18, 21
space race, 5, 22

T

transmitters, 14
Tsiolokovsky, Konstantin, 13

U

United States, 5–6, 18

Web Sites

To learn more about *Sputnik* and other satellites, check out these Web sites:
http://kidsastronomy.about.com
www.yahooligans.com/science_and_nature/the_earth/space/spacecraft/